Information Technology/ Information Management

Strategic Plan
Fiscal Years 2012–2016

AVAILABILITY OF REFERENCE MATERIALS IN NRC PUBLICATIONS

NRC Reference Material

As of November 1999, you may electronically access NUREG-series publications and other NRC records at NRC's Public Electronic Reading Room at http://www.nrc.gov/reading-rm.html.
Publicly released records include, to name a few, NUREG-series publications; *Federal Register* notices; applicant, licensee, and vendor documents and correspondence; NRC correspondence and internal memoranda; bulletins and information notices; inspection and investigative reports; licensee event reports; and Commission papers and their attachments.

NRC publications in the NUREG series, NRC regulations, and *Title 10, Energy*, in the Code of *Federal Regulations* may also be purchased from one of these two sources.
1. The Superintendent of Documents
   U.S. Government Printing Office
   Mail Stop SSOP
   Washington, DC 20402-0001
   Internet: bookstore.gpo.gov
   Telephone: 202-512-1800
   Fax: 202-512-2250
2. The National Technical Information Service
   Springfield, VA 22161-0002
   www.ntis.gov
   1-800-553-6847 or, locally, 703-605-6000

A single copy of each NRC draft report for comment is available free, to the extent of supply, upon written request as follows:
Address: U.S. Nuclear Regulatory Commission
           Office of Administration
           Publications Branch
           Washington, DC 20555-0001
E-mail: DISTRIBUTION.SERVICES@NRC.GOV
Facsimile: 301-415-2289

Some publications in the NUREG series that are posted at NRC's Web site address
http://www.nrc.gov/reading-rm/doc-collections/nuregs
are updated periodically and may differ from the last printed version. Although references to material found on a Web site bear the date the material was accessed, the material available on the date cited may subsequently be removed from the site.

Non-NRC Reference Material

Documents available from public and special technical libraries include all open literature items, such as books, journal articles, and transactions, *Federal Register* notices, Federal and State legislation, and congressional reports. Such documents as theses, dissertations, foreign reports and translations, and non-NRC conference proceedings may be purchased from their sponsoring organization.

Copies of industry codes and standards used in a substantive manner in the NRC regulatory process are maintained at—
   The NRC Technical Library
   Two White Flint North
   11545 Rockville Pike
   Rockville, MD 20852-2738

These standards are available in the library for reference use by the public. Codes and standards are usually copyrighted and may be purchased from the originating organization or, if they are American National Standards, from—
   American National Standards Institute
   11 West 42nd Street
   New York, NY 10036-8002
   www.ansi.org
   212-642-4900

Legally binding regulatory requirements are stated only in laws; NRC regulations; licenses, including technical specifications; or orders, not in NUREG-series publications. The views expressed in contractor-prepared publications in this series are not necessarily those of the NRC.

The NUREG series comprises (1) technical and administrative reports and books prepared by the staff (NUREG-XXXX) or agency contractors (NUREG/CR-XXXX), (2) proceedings of conferences (NUREG/CP-XXXX), (3) reports resulting from international agreements (NUREG/IA-XXXX), (4) brochures (NUREG/BR-XXXX), and (5) compilations of legal decisions and orders of the Commission and Atomic and Safety Licensing Boards and of Directors' decisions under Section 2.206 of NRC's regulations (NUREG-0750).

Message from the Chief Information Officer

I am pleased to issue the U.S. Nuclear Regulatory Commission's (NRC's) Information Technology/Information Management ((IT/IM) Strategic Plan for fiscal years 2012–2016. The mission of the NRC's IT/IM program is to—

Manage information and employ information technology to enhance information access and strengthen agency performance.

By accomplishing the IT/IM mission, we will support the NRC mission to—

License and regulate the Nation's civilian use of byproduct, source, and special nuclear materials in order to protect public health and safety, promote the common defense and security, and protect the environment.

Darren B. Ash

The six outcome goals we have chosen, together with their associated measures and strategies, provide the foundation for directing and assessing the performance of the NRC's IT/IM program over the next 5 years. The plan covers all of the NRC's IT/IM resources agencywide, including our networks, computers, and telecommunications devices; our information and records management; our cyber security operations; and all of our applications, ranging from mission-essential systems, such as those supporting licensing and emergency response, to systems required for support functions such as payroll, personnel, and accounting.

This new edition of the NRC's IT/IM Strategic Plan fully supports the Office of Management and Budget's 25point plan to reform Federal IT, especially in improving IT governance and IT program management. These two areas are critical to ensure that IT investments are properly aligned with agency priorities and provide value, capitalizing on cost-reduction opportunities and mitigating IT risk.

The most important goal of the NRC's IT/IM program is effective information access—enabling both the NRC staff and the public to quickly and easily obtain the information they need. This goal reflects the NRC's commitment to openness and is essential for effective agency operations. The remaining IT/IM goals, covering IT business solutions, IT infrastructure, information and IT security, IT/IM service delivery, and IT/IM skills, all, in one way or another, support the primary goal of effective information access and, in addition, help strengthen agency performance.

The environmental assessment that preceded the development of this plan highlighted the progress the agency has made in such areas as mobile computing, Web-based applications, and IT security, but it also emphasized the challenges we face over the next 5 years with tighter budgets, rapidly changing security threats, and an ever-increasing reliance on technology to accomplish our mission. The assessment also identified some key opportunities, such as system modernization, further improvements in mobile computing, and more effective use of social media and advanced Web technologies, that can help us achieve our goals.

I want to emphasize the importance of performance measurement and accountability in accomplishing the goals of this plan. This strategic plan is not a document to lay aside on the shelf. It is a starting point for our staff to collaboratively set targets, plan the best way to achieve them, and monitor results. In keeping with the Government Performance and Results Modernization Act of 2010, I have committed to conducting quarterly reviews to monitor progress on our goals and strategies, making course corrections as needed. The NRC will report performance for the key IT/IM measures in this plan to Congress, beginning with the results for fiscal year 2013.

These are exciting and challenging times, as we face continued budget constraints and strive to "do more with less" across the Government. However, we have major opportunities to enhance the business of Government and develop a more open and transparent Government through technology. I greatly appreciate the suggestions and comments received from the participants in the IT/IM strategic planning process, and I rely on continued partnership with my fellow executives, managers, and staff across the agency to face these challenges and make the necessary improvements a reality.

Contents

Introduction 4

IT/IM Mission, Vision, and Strategic Goals 5

Strategies, Measures, and Key Activities for Each Goal 5

    Goal 1—Effective Information Access 6

    Goal 2—Effective IT Business Solutions 8

    Goal 3—Effective IT Infrastructure 10

    Goal 4—Secure Information and Information Technology 12

    Goal 5—Service Delivery Excellence 14

    Goal 6—Effective IT/IM Skills 16

APPENDIX A Relationship to the U.S. Nuclear Regulatory Commission Strategic Plan 17

APPENDIX B Planning Approach 20

APPENDIX C Information Technology Business Solutions 22

APPENDIX D Acronyms 23

Introduction

The U.S. Nuclear Regulatory Commission's (NRC's) Information Technology/Information Management (IT/IM) Strategic Plan, Fiscal Years 2012–2016, was prepared under the requirements of the Paperwork Reduction Act and the Clinger-Cohen Act of 1996. This document serves as the NRC's strategic information resources management plan in accordance with Section 3506(b)(2) of the Paperwork Reduction Act.

The NRC used the logic model[1] approach to establish its IT/IM goals and strategies and align them with the broader NRC Strategic Plan. Appendix A provides more information on this alignment.

The planning process emphasized the following:

- beginning with leadership's determination of the underlying outcomes and results expected of IT/IM programs and activities
- facilitating a series of stakeholder sessions on the changes required to meet each strategic goal
- using stakeholder input and broader Federal IT priorities to establish strategies for change
- including performance measures at the goal and strategy levels to ensure results-driven planning
- connecting resources, budgets, and employees with goals and strategies in a structured manner
- providing common definitions and language to assist managers and employees in understanding the connection between their activities and the plan's goals and strategies

Appendix B provides more information on the planning process.

The next sections discuss the following:

- the NRC IT/IM mission, which describes the overall purpose of the IT/IM program
- the IT/IM vision, which describes the desired result of all goals and strategies
- the six goals with their respective outcome measures, the strategies describing the changes needed to achieve each goal, and key activities to implement the strategies

---

[1] W.K. Kellogg Foundation/Mosaica, "Using the Logic Model for Program Planning," http://www.lri.lsc.gov/pdf/other/TIG_Conf._Materials/EMcKay_Logic_Model_Intro_LSC.pdf (accessed April 18, 2011).

IT/IM Mission, Vision, and Strategic Goals

Mission
Manage information and employ information technology to enhance information access and strengthen agency performance.

Vision
Getting the right information to the right people at the right time.

Strategic Goals

Goal 1—Effective Information Access
NRC staff and stakeholders can quickly and easily access the information they need.

Goal 2—Effective IT Business Solutions
The NRC's IT business solutions are easy to use, are cost effective, and strengthen agency performance.

Goal 3—Effective IT Infrastructure
The NRC's IT infrastructure is available, cost effective, and responsive to agency business needs.

Goal 4—Secure Information and Information Technology
The NRC's information and information technology are protected against threats to their integrity, confidentiality, and availability.

Goal 5—Service Delivery Excellence
IT/IM services are readily accessible and effectively delivered to enhance individual and organizational performance.

Goal 6—Effective IT/IM Skills
The NRC workforce has the skills to effectively manage, deliver, and use IT/IM capabilities.

Strategies, Measures, and Key Activities for Each Goal

The following pages lay out the strategies, measures, and key activities for each strategic goal. Each year during the budget process, the NRC will allocate resources for implementing the activities and set the performance targets for each measure. A subset of the measures will be included in the NRC's annual budget transmitted to Congress.

Goal 1—Effective Information Access

NRC staff and stakeholders can quickly and easily access the information they need.

Strategies	Strategy Measures

1. Improve search capabilities.	• Public's score for search effectiveness based on a question from the American Customer Satisfaction Index (ACSI) survey for Federal Web sites
	• Staff's score for search effectiveness based on a question from the IT/IM Employee Survey
	• Percent of identified content sources searchable—
	   ◇ with a single search engine
	   ◇ through the Agencywide Documents Access and Management System (ADAMS) catalog

2. Increase transparency.	• Number of timeliness targets met for key information dissemination channels • Public's score for transparency on the ACSI
3. Improve internal information discovery and usability.	• Score on a standardized set of questions administered to a representative sample of employees measuring their ability to navigate to and use the most commonly accessed NRC content, services, and applications • Staff's score for the effectiveness of the Intranet in helping them obtain and use the information they need to do their jobs
4. Increase efficiency and effectiveness of content dissemination technology.	• Number of publishing platforms and user interfaces for NRC public documents
5. Improve the completeness and accuracy of NRC records and information.	• Number of records classes with retention schedules that are automatically populated in ADAMS Records Manager when an eligible document is declared an official agency record as part of a business process
6. Efficiently provide ongoing information services for staff and the public.	• Percent of service-level targets met

Outcome Measures

- **Public's Score for Information Access:** NRC score on the annual ACSI for Federal Web sites
- **Staff's Score for Information Access:** Average score on the Federal Employee Viewpoint Survey or IT/IM Employee Survey for employees' ability to find and obtain the information they need to do their jobs

Key Activities

- Implement an internal enterprise search to encompass most information in a single search.
- Implement an internal catalog-based search to improve the relevance of search results.

- Employ new Web-based technologies to strengthen user engagement at the public Web site.
- Maintain timely dissemination of key public information, such as public documents and public meeting notices.
- Publish datasets of high value to the public.
- Employ new technologies and best practices to improve public-facing information services.

- Provide an internal user interface that improves the staff's ability to find NRC content, services, and applications.
- Develop and execute a plan to improve the management and utility of the agency's intranet and internal collaboration tools.

- Implement a single publishing platform for NRC content in the Digital Document Management System, the Electronic Hearing Docket (EHD) and the Publicly Available Records System (PARS).
- Implement a single public Web user interface to facilitate access to documents in PARS and EHD.
- Implement a content management system to improve the efficiency of content publication on the agency's public Web site.

- Modernize records and information management (RIM) processes to make information capture and categorization more complete and transparent.
- Develop new file plans, records categorization, and retention schedules for use in records capture.
- Improve the staff's knowledge of RIM related to NRC work products.
- Incorporate RIM policies in the NRC's capital planning and investment control process and in the agency's IT project management methodology (PMM).

- Deliver basic information services.
- Respond to Federal mandates related to information management.
- Maintain vendor support for content management systems such as ADAMS and NRC Web sites.
- Seek efficiencies in information service operations and make improvements.

Goal 2—Effective IT Business Solutions

The NRC's IT business solutions[1] are easy to use, are cost effective, and strengthen agency performance.

Strategies	Strategy Measures
1. **Strengthen enterprisewide planning and standards.**	• Percent of IT investments approved through the agency's IT capital planning and investment control process that conform to agency's investment planning and submittal requirements
2. **Strengthen planning for NRC business area processes and their supporting IT.**	• Percent of IT projects that are part of an approved business area plan
3. **Strengthen project selection, execution, and control.**	• Governance effectiveness score based on a survey of the participants • Average postimplementation evaluation score for completed projects
4. **Efficiently provide ongoing IT business solution services.**	• Percent of service-level targets met

---

1   See Appendix C

Outcome Measures

- **IT Investment Management Score:** Average score on a scale of 1 to 10 for NRC IT investments reported to the Office of Management and Budget
- **Staff's Score for IT Business Solutions:** Average score for two questions on the Federal Employee Viewpoint Survey or IT/IM Employee Survey regarding the effectiveness of IT business solutions

Key Activities

- Update IT system inventory information and improve information quality.
- Provide IT architecture requirements and guidance.
- Annually update the agency enterprise architecture transition plan.
- Provide enterprisewide IT contracts that facilitate acquisition of IT resources that conform to agency standards, architecture, and guidance.

- Establish a governance framework for IT business area planning.
- Assess the current planning state of each NRC business area and establish baseline metrics that can be used for measuring and influencing improvements.
- Develop business area plans that include a sequence of IT projects to be submitted to the agency's IT capital planning process.
- Standardize agencywide processes and associated IT solutions for support functions such as financial management, human resources management, and administrative services.

- Partner with procurement officials to ensure that IT purchases conform to agency policies and standards.
- Revise the NRC's investment management measure criteria to motivate improvements in key areas.
- Complete postimplementation evaluations for each project within 1 year from the start of production operations.
- Institutionalize and communicate lessons learned from postimplementation reviews.

- Deliver basic IT applications infrastructure services, such as database administration services and enterprise contracts for systems development and maintenance.
- Respond to Federal mandates related to the applications infrastructure and IT investment governance.
- Maintain vendor support for IT applications infrastructure hardware, software, and operating systems.
- Seek efficiencies in IT business solution service operations and make improvements.

Goal 3—Effective IT Infrastructure

The NRC's IT infrastructure is available, cost effective, and responsive to agency business needs.

Strategies	Strategy Measures
1. Expand tools and services for "working from anywhere."	• Staff's score for availability of the IT tools and support services needed to work effectively, regardless of physical location • Percent of NRC staff whose office-defined mobile device requirements are met • Increase in the number of user-identified high-priority functions made available for access on mobile devices
2. Expand tools and services for "working with anyone."	• Staff's score for the availability of IT tools and support services needed to meet and collaborate effectively both internally and with external stakeholders • Increase in the annual number of virtual public meetings held, including both Web streaming and Web conferencing
3. Streamline the sign-on process.	• Percent of NRC-controlled systems requiring access controls that use the NRC LAN account or NRC Personal Identity Verification (PIV) card as the means to control user access
4. Strengthen business continuity capabilities.	• Number of high-priority applications that have been successfully tested for their ability to perform the required functions after a disruptive event
5. Efficiently provide ongoing infrastructure services.	• Percent of service-level targets met • Percent of identified infrastructure components that are supported by the current vendor

Outcome Measures

- **Availability:** Availability of key network services
- **Staff's Score for IT Infrastructure:** Average score on the Federal Employee Viewpoint Survey or IT/IM Employee Survey regarding whether the agency provides the IT infrastructure the staff needs to work effectively

Key Activities

- Develop and implement an agencywide mobile computing strategy, including devices (Government-supplied and personal), infrastructure, applications, and security.
- Expand mobile and remote access to agency applications.
- Implement a communications plan to encourage application sponsors to make their applications useful on portable devices using enterprise contracts and standards.
- Enhance mobile device capabilities and support to the staff.

- Expand awareness of virtual meeting capabilities.
- Expand NRC's official presence on social media based on business value.
- Expand capabilities and support to facilitate information sharing, collaborative development of work products, and use of collaboration Web sites, both internally and with external stakeholders.

- Provide and support an agencywide sign-on infrastructure so that new and existing applications will not need a separate user identification and password and can accept an NRC badge (PIV card) for access.
- Implement policies to ensure that new systems or major enhancements use the agency's sign-on infrastructure.

- Validate the list of high-priority applications.
- Test and strengthen contingency plans.
- Support the agency disaster recovery program by identifying and meeting requirements related to high-priority applications.

- Deliver basic infrastructure services.
- Respond to Federal mandates related to the IT infrastructure.
- Maintain vendor support for infrastructure hardware, software, and operating systems.
- Extend all relevant infrastructure services to the NRC's new building and make infrastructure changes necessary to consolidate existing offices.
- Provide information to help program managers (PMs) and the Information Technology Business Council better estimate infrastructure costs.
- Seek efficiencies in IT infrastructure service operations and make improvements.

Goal 4 — Secure Information and Information Technology

The NRC's information and information technology are protected against threats to their integrity, confidentiality, and availability.

Strategies	Strategy Measures
1. **Implement information security policies that balance security risk with operational mission needs.**	• Percent of operational technologies for which a risk-informed configuration standard has been issued according to the configuration standard development plan • Percent of policy updates for which either a revision or a plan for updating the policy is issued within agreed-upon timeframe
2. **Reduce information security risks resulting from human actions.**	• Percent of network users who were susceptible to simulated malware e-mails • Percent reduction in the number of information security infractions, both cyber and nonelectronic
3. **Reduce IT system security risks.**	• Percent of IT projects and operational systems that have a current authority to operate from the designated approving authority • Percent of operational applications and general support systems that met the NRC's annual risk management activities requirements in accordance with guidance from the Chief Information Officer
4. **Reduce operational cyber security risks.**	• Reduction in the number of repeat findings in network penetration testing • Percent of controls implemented according to the agency's Consensus Audit Guideline Plan

Outcome Measures

- **Security Incidents:** Number of IT security incidents reported to the Department of Homeland Security that significantly affect the NRC's access to information or IT services.
- **Cyber Security Program Effectiveness:** Rating of the NRC's cyber security program effectiveness based on the Inspector General's annual Federal Information Security Management Act audit.

Key Activities

- Implement the Federal Government's controlled unclassified information program.
- Review and update the NRC information security policy, standards, guidelines, training, and contract language, as needed, to reflect Federal information security and NRC operational mission requirements.
- Partner with procurement officials to incorporate current information security requirements in NRC IT contracts.
- Coordinate and communicate information security policy changes with affected NRC organizations.
- Determine exceptions to cyber security policies by weighing risks versus business needs.

- Provide information security awareness training and role-based training that address the current threat environment.
- Conduct social engineering tests and provide real-time security training to affected employees.
- Monitor information security infraction and incident trends and take actions to address issues.

- Automate and maintain a system authorization process to help reduce cost and complexity.
- Automate and maintain the Plan of Actions and Milestones (POA&M) process to effectively manage remediation of system-level vulnerabilities.
- Develop an enterprise risk management plan in accordance with National Institute of Standards and Technology Special Publication 800 39, "Managing Information Security Risk."
- Develop and maintain a plan for identifying and monitoring common controls agencywide.
- Streamline the agency's continuous monitoring process to minimize manual reviews.

- Conduct independent penetration testing.
- Mitigate risks identified in POA&Ms of IT infrastructure systems.
- Implement the critical controls from the Federally accepted consensus audit guidelines.
- Provide a centralized repository of information for cyber security incident management.
- Identify and implement better tools and processes to detect, mitigate, and contain damage from cyber security incidents.
- Assess requirements and fill IT security capability gaps to support endpoint situational awareness.
- Implement a secure means of communicating sensitive information with key external stakeholders.

Goal 5—Service Delivery Excellence

IT/IM services are readily accessible and effectively delivered to enhance individual and organizational performance.

Strategies	Strategy Measures
1. Adjust IT/IM services and service levels based on customer feedback and resource constraints.	• Number of service areas with at least one service level established
2. Improve the service request and service delivery process.	• Number of requests that come through the new service delivery channel framework
3. Improve efficiency of service delivery.	• Cost of identified IT/IM services
4. Increase awareness and enhance communication of IT/IM services.	• Staff's score for IT/IM communications based on a question on the IT/IM Employee Survey

Outcome Measures

- **Service-Level Targets Met:** Percent of service-level targets met
- **Staff's Score for IT/IM Services:** Average score on the Federal Employee Viewpoint Survey or IT/IM Employee Survey regarding whether the agency provides the IT/IM support services the staff needs to work effectively

Key Activities

- Complete gap analysis for IT/IM services and service levels and make necessary adjustments.
- Identify and address needs of offices without dedicated IT staff.
- Periodically review performance with customers and make adjustments to the services, service levels, or service level agreements, as appropriate.
- Adjust services and service levels in response to resource constraints.

- Design a conceptual IT/IM service delivery channel framework.
- Implement a detailed IT/IM service delivery channel operational architecture.
- Develop a project services workload management framework.

- Establish the baseline cost of identified IT/IM services.
- Develop cost-analysis framework and approach for IT/IM services and pilot the selected approach on two services.
- Compare cost data for the first two services against industry and government benchmarks.
- Repeat the process for the rest of the identified IT/IM services.
- Sustain the IT/IM services cost model and link to IT/IM priority workload management and governance.
- Foster a partnering and collaborative environment for service providers across the Office of Information Services organization.

- Develop and use a consistent strategic framework for IT/IM communications.
- Measure and publish performance against validated service levels or service level agreements.
- Implement independent recommendations for enhanced IT/IM communications.
- Update and execute communications plans for IT/IM services.

Goal 6 – Effective IT/IM Skills

The NRC workforce has the skills to effectively manage, deliver, and use IT/IM capabilities.

Outcome Measure

- **IT/IM Competency:** Average supervisor evaluation of staff competency in selected IT/IM skill areas.

Strategies	Strategy Measures	Key Activities
1. **Institute competency-based training and qualification for key IT/IM roles.**	• Percent of employees in identified roles that have completed their qualifications programs	• Establish a governance structure to oversee training for IT/IM professionals and identify high priority roles, such as IT PMs and information system security officers. • Develop and implement a qualifications program for each role.
2. **Improve employee training for new IT capabilities.**	• Training evaluation scores for new IT capabilities	• In partnership with NRC Human Resource (HR) staff, review and update the PMM and the criteria used in project control reviews to improve the planning and development of training associated with major new IT capabilities. • In partnership with HR, periodically communicate the agency's IT training guidance and best practices to IT PMs.
3. **Improve recruitment of IT/IM talent.**	• Number of staff recruited and selected from recruitment events attended	• Participate in internal and external events for targeted recruitment. • Recruit participants for the Student Career Experience Program from IT job shadow day.

APPENDIX A
Relationship to the U.S. Nuclear Regulatory Commission Strategic Plan

This appendix shows the relationship between specific strategies, means, and activities for the goals and organizational excellence objectives in the U.S. Nuclear Regulatory Commission's (NRC's) Strategic Plan for fiscal years (FYs) 2008–2013 and the six information technology/information management (IT/IM) strategic goals.

The U.S. Nuclear Regulatory Commission Strategic Plan

Mission

License and regulate the Nation's civilian use of byproduct, source, and special nuclear materials to ensure adequate protection of public health and safety, promote the common defense and security, and protect the environment.

Strategic Goals

Safety: Ensure adequate protection of public health and safety and the environment.

Security: Ensure adequate protection in the secure use and management of radioactive materials.

Strategic Outcomes

Safety:
- Prevent the occurrence of any nuclear reactor accidents.
- Prevent the occurrence of any inadvertent criticality events.
- Prevent the occurrence of any acute radiation exposures resulting in fatalities.
- Prevent the occurrence of any releases of radioactive materials that result in significant radiation exposures.
- Prevent the occurrence of any releases of radioactive materials that cause significant adverse environmental impacts.

Security:
- Prevent any instances where licensed radioactive materials are used domestically in a manner hostile to the United States.

Organizational Excellence Objectives

Openness: The NRC appropriately informs and involves stakeholders in the regulatory process.

Effectiveness: NRC actions are of high quality, efficient, timely, and realistic, to enable the safe and beneficial uses of radioactive materials.

Operational Excellence: NRC operations use effective business methods and solutions to achieve excellence in accomplishing the agency's mission.

(The full text of the plan may be found at: http://www.nrc.gov/reading-rm/doc-collections/nuregs/staff/sr1614/v4/index.html)

Linkage to NRC Strategic Plan

The numeric entries under each IT/IM goal indicate an associated IT/IM strategy that supports the corresponding element of the NRC Strategic Plan.

Goals and Selected Strategies, Means, and Activities in the NRC FY 2008–2013 Strategic Plan	IT/IM Goals with Relevant Strategies					
	Information Access	IT Business Solutions	IT Infrastructure	Secure Info and IT	Service Delivery Excellence	IT/IM Skills
SAFETY GOAL						
Continue to oversee the safe operation of existing plants while preparing for and managing the review of applications for new power reactors. (Strategy 2)	1, 2, 5, 6	3, 4	5			
Effectively respond to events at NRC-licensed facilities and other events of national interest, including maintaining and enhancing the NRC's critical incident response and communication capabilities. (Strategy 9)		5	1, 5			
SECURITY GOAL						
Use relevant intelligence information and security assessments to maintain realistic and effective security requirements and mitigation measures. (Strategy 1)				1, 4		
Control the handling and storage of sensitive security information and the communication of information to licensees and Federal, State, and local partners. (Strategy 4)				2, 3		
Enhance the programs for control of the security of radioactive sources and strategic special nuclear material commensurate with their risk, including enhancements required by the Energy Policy Act of 2005. (Strategy 7)		3		3		
Collaborate with the U.S. Department of Energy, U.S. Department of Homeland Security, and other agencies and State governments to develop and implement a national registry of radioactive sources. Improve the controls on high-risk radioactive materials, including enhancements required by the Energy Policy Act of 2005 and recommended by the Task Force on Radiation Source Protection and Security, to prevent their harmful use. (Means supporting Strategies 1, 4, 6, and 7)		3		3		
Identify and develop key information technology investments, including secure electronic document and records management capabilities, that will enhance the storage, handling, and communication of sensitive security information both within and external to the agency. (Means supporting Strategy 4)		3		3		

Goals and Selected Strategies, Means, and Activities in the NRC FY 2008–2013 Strategic Plan	IT/IM Goals with Relevant Strategies					
	Information Access	IT Business Solutions	IT Infrastructure	Secure Info and IT	Service Delivery Excellence	IT/IM Skills
OPENNESS (Organizational Excellence Objective)						
Enhance the awareness of the NRC's independent role in protecting public health and safety, the environment, and the common defense and security. (Strategy 1)	3, 4, 5, 6					
Provide accurate and timely information to the public about NRC's mission, regulatory activities, and performance and about the uses of, and risks associated with, radioactive materials. (Strategy 2)	3, 4, 5, 6		5			
Provide for fair, timely, and meaningful stakeholder involvement in NRC decisionmaking without disclosing classified, safeguards, proprietary, and sensitive unclassified information. (Strategy 3)	3, 4, 5, 6			2, 3		
EFFECTIVENESS (Organizational Excellence Objective)						
Continue to improve the NRC's regulatory and communication programs. (Strategy 7)		2, 3				
Achieve efficiencies in the licensing process that enable the safe and secure use of nuclear material. (Strategy 8)		2, 3				
OPERATIONAL EXCELLENCE (Organizational Excellence Objective)						
Improve support services to make them more efficient and make it easier to accomplish agency goals. (Strategy 3)					1, 2, 3, 4	
Manage information and employ information technology to improve the productivity, effectiveness, and efficiency of agency programs and enhance the availability and usefulness of information to all users inside and outside the agency. (Strategy 4)	All	All	All	All	All	All
Use innovative strategies to recruit, develop, and retain a high-quality, diverse workforce. (Strategy 5)						1, 2
Sustain a learning environment that provides continuing improvement in performance through knowledge management, performance feedback, training, coaching, and mentoring. (Strategy 7)						1, 2
Provide accurate, timely, and useful financial information to agency managers for effective decisionmaking. (Strategy 9)		2, 3				

APPENDIX B
Planning Approach

To develop this plan, the U.S. Nuclear Regulatory Commission (NRC) used a proven methodology, based on the logic model, that resulted in measurable goals and strategies intended to drive performance improvement over the next several years.

The pyramid in Figure 1 depicts the outcomes of the planning process in the top half and the relationship to other plans in the bottom half.

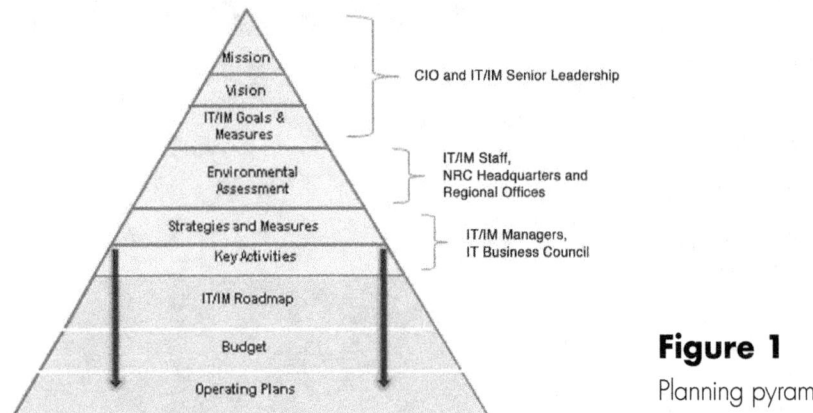

Figure 1
Planning pyramid

To begin the information technology/information management (IT/IM) strategic planning process, the NRC's Chief Information Officer (CIO), together with other IT/IM senior executives, agreed upon the IT/IM program mission (based on language adopted from the NRC Strategic Plan), vision, and six measurable goals. These measurable outcomes were designed to provide 3- to 5 year indicators of success for all IT/IM employees and internal and external users of NRC information.

The six goals established by the IT/IM senior leadership were then used individually as the desired result for a series of strategic environmental assessments. These sessions, each with a specifically chosen group of NRC IT/IM internal stakeholders, used the goals to drive a structured, facilitated process to derive a comprehensive list of internal and external issues, in both present and expected future operating environments.

The results of the environmental assessment were used as input for the development of strategies, strategy measures, and key activities. Strategies are defined as changes the agency must make to accomplish the goals. Strategy measures monitor the progress made on the strategies and serve as leading performance indicators for the goal measures. Key activities are the projects or initiatives the agency will undertake to implement each strategy.

As shown in the bottom half of the pyramid, the agreed-upon activities will be used to drive the agency's IT/IM Roadmap (a more tactical plan for setting technology direction), the IT/IM budget, and the annual operating plans for the major IT/IM organizations.

The NRC IT/IM Strategic Plan was developed in close conjunction with the new Governmentwide plan to reform Federal IT management. This plan and related Governmentwide guidance underscore the governance and management concerns that are an issue for both NRC and all Federal technology investments. Congress, the Government Accountability Office, and the Office of Management and Budget have repeatedly pointed to IT investments that often cost more, take longer, and deliver less than promised. With a large portion of Federal discretionary funding invested in IT and IM, the Federal CIO has made a 25 point, 18 month plan a Governmentwide priority.

This strategic plan follows the Federal CIO's guidance by achieving the following:

- The Information Technology Business Council has been strengthened, as have the use of enhanced data and increased early transparency into poorly performing projects. Stronger agencywide IT governance will focus on rapid course correction and timely decisions to change scope before poorly performing investments consume additional resources.

- The use of cloud computing will be increased to improve the return on IT investments and decrease the total cost of ownership. New software solutions will favor software as a service model that requires no additional NRC footprint.

- Strengthened governance and funding models will ensure centralized planning, lower cost per user, and leveraging of expertise from the Office of Information Services and the CIO in all agency IT acquisitions and IM investments.

- Stronger performance measures will be developed and used to accurately gauge IT performance, and the performance review process will be improved. These changes will allow for a much faster response to project impediments or requirement changes.

APPENDIX C
Information Technology Business Solutions

To categorize its information technology (IT) business solutions (referred to here as IT investments), the U.S. Nuclear Regulatory Commission (NRC) uses the Office of Management and Budget (OMB) segment architecture approach. This provides a framework and vocabulary common across the Federal Government. As shown in Figure 2 below, the segments are organized into three groups: core mission areas; business services, and enterprise services. As illustrated in the figure, enterprise services support both core mission areas and business services. These categories support analysis of IT investments to ensure optimal use of resources.

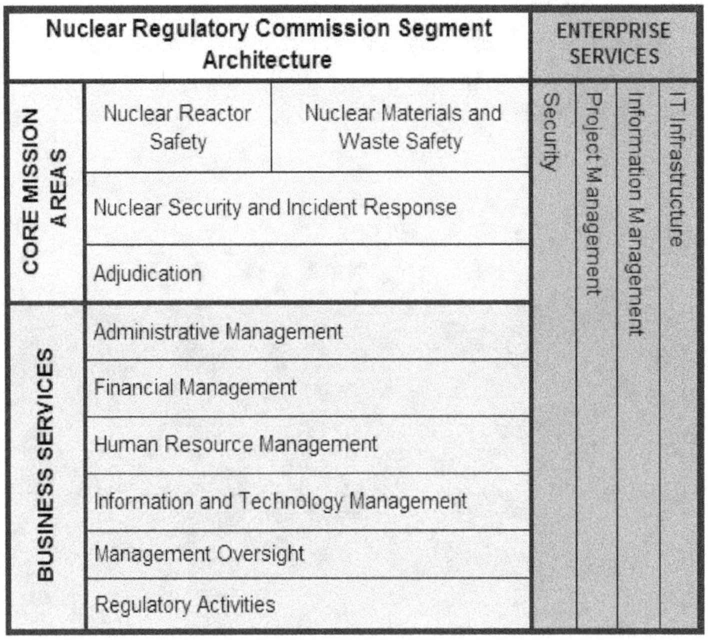

Nuclear Regulatory Commission Segment Architecture			ENTERPRISE SERVICES			
CORE MISSION AREAS	Nuclear Reactor Safety	Nuclear Materials and Waste Safety	Security	Project Management	Information Management	IT Infrastructure
	Nuclear Security and Incident Response					
	Adjudication					
BUSINESS SERVICES	Administrative Management					
	Financial Management					
	Human Resource Management					
	Information and Technology Management					
	Management Oversight					
	Regulatory Activities					

Figure 2
NRC segments

The Federal IT Dashboard is a Website enabling Federal agencies, industry, the general public, and other stakeholders to view details of Federal IT investments. Information on specific NRC investments can be found within the NRC Porfolio on this Website. Select the "Investments" tab to view the list of investments.

APPENDIX D
Acronyms

Acronym	Definition
ADAMS	Agencywide Documents Access and Management System
ACSI	American Customer Satisfaction Index
C&A	certification and accreditation
CIO	chief information officer
DDMS	Digital Data Management System
ECM	enterprise content management
EHD	Electronic Hearing Docket
FY	fiscal year
HR	human resources
HSPD-12	Homeland Security Presidential Directive 12
ID	identification
IM	information management
ISSO	information systems security officer
IT	information technology
ITSAC	Information Technology Senior Advisory Council
NRC	U.S. Nuclear Regulatory Commission
PARS	Publicly Available Records System
PIV	personal identity verification
PM	program manager
PMM	project management methodology
POA&M	plan of action and milestones
RIM	records and information management

NRC FORM 335 (12-2010) NRCMD 3.7	U.S. NUCLEAR REGULATORY COMMISSION BIBLIOGRAPHIC DATA SHEET *(See instructions on the reverse)*	1. REPORT NUMBER (Assigned by NRC, Add Vol., Supp., Rev., and Addendum Numbers, if any.) NUREG-1908, Vol. 2
2. TITLE AND SUBTITLE United States Nuclear Regulatory Commission Information Technology/Information Management Strategic Plan Fiscal Years 2012-2016		3. DATE REPORT PUBLISHED MONTH: 12 YEAR: 2011
		4. FIN OR GRANT NUMBER
5. AUTHOR(S) Francine F. Goldberg Senior Level Advisory for IT/IM Strategic Planning and Communication		6. TYPE OF REPORT
		7. PERIOD COVERED (Inclusive Dates)

8. PERFORMING ORGANIZATION - NAME AND ADDRESS (If NRC, provide Division, Office or Region, U. S. Nuclear Regulatory Commission, and mailing address; if contractor, provide name and mailing address.)
Office of Information Services
U.S. Nuclear Regulatory Commission
Mail Stop O6E3M
Washington, D.C. 20555

9. SPONSORING ORGANIZATION - NAME AND ADDRESS (If NRC, type "Same as above", if contractor, provide NRC Division, Office or Region, U. S. Nuclear Regulatory Commission, and mailing address.)
Same as above

10. SUPPLEMENTARY NOTES

11. ABSTRACT (200 words or less)
The U.S. Nuclear Regulatory Commission's Information Technology/Information Management (IT/IM) Strategic Plan for Fiscal Years 2012-2016 describes how IT/IM activities at the NRC helps accomplish the agency's mission. The IT/IM plan responds to Federal requirements in the Paperwork Reduction Act (PRA) and the Clinger-Cohen Act of 1996, serving as the NRC's strategic information resources management plan in accordance with Section 3506(b)(2) of the PRA. The plan lays out the mission and vision for the agency's IT/IM programs and establishes six goals along with strategies for accomplishing them. It also defines measures of success in attaining the goals. These goals, strategies, and measures provide the foundation for directing and assessing the performance and results of the NRC's IT/IM over the next 3-5 years.

12. KEY WORDS/DESCRIPTORS (List words or phrases that will assist researchers in locating the report.) strategic plan, information technology, information management, goals, strategies, strategic IRM plan, IT, IT applications, IT infrastructure, IT security	13. AVAILABILITY STATEMENT unlimited
	14. SECURITY CLASSIFICATION *(This Page)* unclassified *(This Report)* unclassified
	15. NUMBER OF PAGES 28
	16. PRICE

NUREG-1908, Vol. 2
December 2011